GW00706041

A Science Museum illustrated booklet

MICROSCOPES

to the end of the nineteenth century

by F. W. Palmer
and A. B. Sahiar

Her Majesty's Stationery
Office London 1971

SBN 11 290105 0

Cover: Hooke's microscope (c1665) see plate 3

Note: All the items illustrated in this booklet are part of the Science Museum collections

Introduction

Between Seneca's globe of water (c100 AD) through which 'letters though small and indistinct are seen enlarged and more distinct', and the modern electron microscope, is a fascinating story of man's endeavour to see more than nature intended him to see with the unaided eye.

It is conceivable that the magnifying power of transparent media with convex surfaces was known very early, possibly even to the Greeks and the Romans. It has been argued that the minuteness of some of the ancient pieces of workmanship may indicate the use of 'Magnifying Glasses' by the artisans. The first illustration showing the use of magnifying glass was made by George Hoefnagel in 1592. The engraved copper plates of his work illustrate ordinary natural objects with minute accuracy and showing details which would hardly be distinguishable to the naked eye. An Arabian Scholar Al Hazen (Ibn-al-Haitham c1000) is reputed to be the first to have understood the working of the human eye; the latin translations of his writings also indicate his keen interest in the theory of lenses.

The next step seems to have been the use of a pair of lenses, one for each eye; The Spectacles. The inventor is not known with any certainty but it could have been any worker in glass near the end of the thirteenth century. The Chinese are reputed to have already had reading glasses.

Antony van Leeuwenhoek (1632-1723) from Delft, was the first man to show what a single lens, in the hands of a man who had the art of grinding and polishing it, could do to reveal the secrets of nature. With one of his home-made lenses he attained a magnification of x270. It is with these 'Simple Microscopes' that he astonished the philosophers of the period with his discoveries of infusoria, bacteria and other microscopic forms of life. He published numerous letters in the Philosophical Transactions of the Royal Society of London on his observations and discoveries. He was elected a fellow of the Royal Society in 1680.

The other important Simple Microscopes were the instruments made by Musschenbroek (1660-1707) of Leyden, the compass microscope, Wilson's screwbarrel (1702) microscope, and the dissecting microscope (c1750) by Lyonet. The compass and the screwbarrel were so popular as portable microscopes that they were made throughout the 18th Century.

It is worth noting here that although it has been referred to in the literature by his name, Wilson never claimed it as his own invention; Nicolass Hartsoeker (c1694) has been credited to have been the first to use this type of screw-barrel focussing for a simple microscope. The 'Lieberkühn' which is attached to most of the late forms of these microscopes was first made in 1738 by a German called Lieberkühn. Described by Descartes in 1637, it is essentially a silvered concave reflector with a hole cut out in the centre for the lens. It focusses the light on an opaque object viewed through the lens.

Although the Compound Microscope and the Simple Microscope developed initially side by side we have carried on describing the Simple Microscopes simply to keep the continuity of thought. There is no easy answer to the question, 'Who made the first Compound Microscope?' The early history of this instrument is rather muddled. Cornelius Drebbel is reputed to have demonstrated a compound microscope in London in 1621. This instrument was later carried to Paris and Rome; an instrument presumably made in Holland. Early

European instrument makers like Wiesel, Divini, Campani and Borel are worth noting, but one of the landmarks in the history of the compound microscope was the publication of *Micrographia* by Robert Hooke in 1665.

His early apprenticeship to a portrait painter coupled to his training at Oxford helped him to produce such detailed and accurate illustrations of the various objects he observed under the microscope. On examining the smooth and polished point of a needle under the microscope he writes 'not round nor flat, but irregular and uneven; so that it seem'd to have been big enough to have afforded a hundred armed Mites room enough to be rang'd by each other without endangering the breaking one anothers necks, by being thrust off on either side. The surface of which, though appearing to the naked eye very smooth, could not nevertheless hide a multitude of holes and scratches and ruggedness from being discover'd by the Microscope'. About the natural objects observed under the microscope, '. . . whereas in the works of Nature, the deepest Discoveries show us the greatest Excellencies'.

The 'optics' of his microscope namely the objective, the 'field-lens' and the eye-lens remained as the basic optical system of the English microscopes for almost a hundred years. His microscope was obviously designed primarily for viewing opaque objects, impaled on the point of a pin; the instrument would have to be inclined horizontal to view transparent objects.

Twenty-five years after the publication of his *Micrographia*, Hooke was complaining that only one man was working on microscopy, namely Leeuwenhoek, the rest were merely using the instrument for 'diversion and pastime'. This emphasis on the amateur being the main patron of the microscope makers in England is borne out by the fact that the next hundred years were devoted to making beautiful instruments, more and more convenient to use mechanically but with no attempts to make real advances on the optical design.

The famous instrument makers of the eighteenth century concentrated on improvements, to the stand of the microscope, the various types of illumination systems, the stage holding the specimen and the various arrangements for focussing the instrument.
Flicking through the photographs one can see the gradual but logical development of the stand towards the modern optical microscopes; from the rather flimsy wooden bases and metal rod pillars to the more sturdy and fairly stable later models, sometimes ornate and elaborate but always moving towards greater stability.
In the early days of microscopy the specimens were opaque and were examined under a microscope by sunlight or the light of a lamp, reflected from the object. Hooke focussed the light from an oil lamp on the object with the help of a glass sphere filled with water and a condenser lens. Microscope stages even as late as the nineteenth century had provision for mounting a condenser lens on top of the stage. Later, with the growing need for observing transparent objects, plane mirrors, concave mirrors, lenses and systems of condensers with diaphragms were mounted below the stage; a mirror reflecting the light from a source upwards into the condenser system.
Once again, for the stage of the microscope, one notices the logical development from the early cumbersome stages when the pillars holding the stage or the various attachments got in the way of easy operations on the stage, to the flat clear stages which could be rotated or moved in any direction to facilitate the selection of various parts of an object for observation. Some stages of the late eighteenth and the early nineteenth centuries were fascinatingly elaborate; providing attachments for holding frogs or fishes to examine the circulation of the blood.
The specimens were mounted on ivory slides in which holes were recessed, the object being sandwiched between two mica discs called 'talcs' slipped into the recesses and held together by a brass

ring. This method of mounting specimens was used even up to the middle of the nineteenth century; the common objects being butterfly scales, animal hairs, wings of flies, scales of fishes, insects, the most popular being the flea and the louse from the human head.

The various fundamental mechanical ways of altering the distance between any two objects, have been used for the focussing systems of the microscopes; the two arms of a compass hinged at one end and swinging in and out at the other end, tubes sliding into one another, tubes screwing into one another, a tube sliding over a rod, or a nut moving along the length of a screw and the rack and pinion.

The sliding tubes were used (c1667) by the Italian maker Divini for his first microscopes while his compatriot Campani has used tubes threaded into one another. Hooke probably was the first (c1665) to use a tubular piece sliding along a rod for coarse focussing. Hevelius in 1673 described a fine-focussing arrangement which in an improved form was used extensively by Marshall (c1695) on his microscopes; a long metal screw moving the microscope body up and down as the nut is turned. Bonanni as early as 1691 had used a rack and pinion arrangement for coarse focussing of his horizontal instrument. Microscope makers of the later centuries have merely improved on the precision and the mechanical construction, using the same basic ideas of the 17th Century. Some have combined more than one of the above systems in the same instrument while others have varied the moving parts of the instruments, for instance moving the stage up and down instead of the tube, to focus the microscope.

Most of the English microscopes, up to the middle of the 18th century had the standard 'three lens' optical system of the period namely the objective, the field-lens and an eye-lens. It was realised fairly early in the development of the compound microscope that although one could magnify objects with just two convex lenses, the field of view obtained with this system was very small indeed, hence the introduction (c1654) of a field-lens between the objective and the eye-lens.

The two major shortcomings of a lens, namely the spherical aberration and the chromatic aberration, must have exasperated the first serious research workers with these microscopes. The spherical aberration in a thick lens causes blurring of the image because the image is drawn out along the axis of the lens; the rays from the periphery of the lens forming an image nearer the lens while the rays through the centre of the lens form an image slightly further away from the lens along the axis. The early microscope makers had understood this problem and had circumvented it by putting a small diaphragm behind the objective, thus cutting out the peripheral rays and using only the narrow beam through the centre of the lens. This naturally made the image sharper but, less bright, hence the development of various forms of substage illumination to increase the brightness of the image.

Chromatic aberration, due to the various colours of the spectrum forming their individual images at different points along the axis of a lens, defied solution for more then a century. Even twenty five years after Dollond put on the market (1758) his achromatic combination for telescope lenses, the lens grinders of the country were still struggling to make a tiny microscope objective achromatic.

Van Deijl of Amsterdam put achromatic objectives on the market by 1807. He used a suitable combination of two lenses one of crown glass and the other of flint. At about this time in France, Chevalier was experimenting with cemented doublets and Selligue was trying to make (c1824) a high power achromatic objective by screwing together several low power (more easily ground) achromatic lens combinations. Lister, father of the famous surgeon, showed, (c1830) that although this screwed-in lens combination was achromatic the 'spherical errors' were accumulated and rendered the combination useless for work where good resolution was required. While the European workers were struggling empirically to correct for both types of errors, Lister gave in his paper (Philosophical Transactions, 1830) the theoretical lead to all microscopists, to design their lens combinations rather than make them by trial and error. Ross made lenses to Lister's design and by the middle of the 19th century his lenses had become popular. Ross in conjunction with Wenham also introduced (c1860) the binocular microscopes in this country.

Amici in Italy was the first (c1850) to popularise the use of an immersion objective, a principle which later proved to be the main contribution towards the attainment of the limit of optical resolution. The microscopist Stephenson suggested (c1878) to the famous German physicist Ernst Abbé the idea of homogeneous immersion. In partnership with Carl Zeiss at Jena, Abbé brought on the market the oil immersion objectives which proved extremely popular especially to bacteriologists; being their most fruitful period at the time.

These above mentioned defects in the optical system of the early microscope, up to the middle of the 19th century, confused and misled the microscopists of the period; even experienced workers like Leeuwenhoek, Malpighi, Swammerdam etc., drew conclusions from their observations which later proved to be false. Artifacts, due to the various types of distortions, were taken as new types of structures of the organisms or parts of anatomy examined under the microscope. However by the end of the 19th century the techniques of the microscope had reached a stage of perfection which helped biologists and men of medicine to make discoveries that have contributed so much towards the alleviation of human suffering.

1 Early simple microscopes

At the top of the picture is a copy of an original simple microscope made by Leeuwenhoek (c1673) preserved at the University Museum of Utrecht. Anthony van Leeuwenhoek was born in Delft in 1632. Trained as a linen draper in Amsterdam he returned to Delft in 1654 to establish his own business. Apart from being a successful businessman he became a qualified surveyor in 1669 and was appointed a wine gauger in 1679. Thanks to his numerous published papers relating to his discoveries with his home-made microscopes, he was made a fellow of the Royal Society, London in 1680. His microscope in the Utrecht museum essentially consists of a single double-convex lens with a magnification of x275, mounted between two brass plates rivetted together. The long thumb-screw on a swivelling angle-piece positions the object on the point of the small rod mounted on a brass block. The short thumb-screw moves this block against the plate to focus the object.
The microscope on the right was made by Musschenbroek in Leyden (c1700). Johan Joosten van Musschenbroek was born in Leyden in 1660 in a family interested in scientific apparatus. Musschenbroek became famous in his time as a maker of two well-known types of simple microscopes. The main feature of his microscope as illustrated here, is the use of ball and socket joints, for the first time in the development of microscopes, to facilitate the adjustment of the position of the object in front of the lens. He used ground bi-convex lenses or glass spheres.
The ivory handle microscope was made by Depovilly (c1660-1710) in Paris. A multiple object-disc rotates between two plates hinged at the handle. One of these plates carries the lens holder and the other the object-disc. The notched head screw is used for focussing.

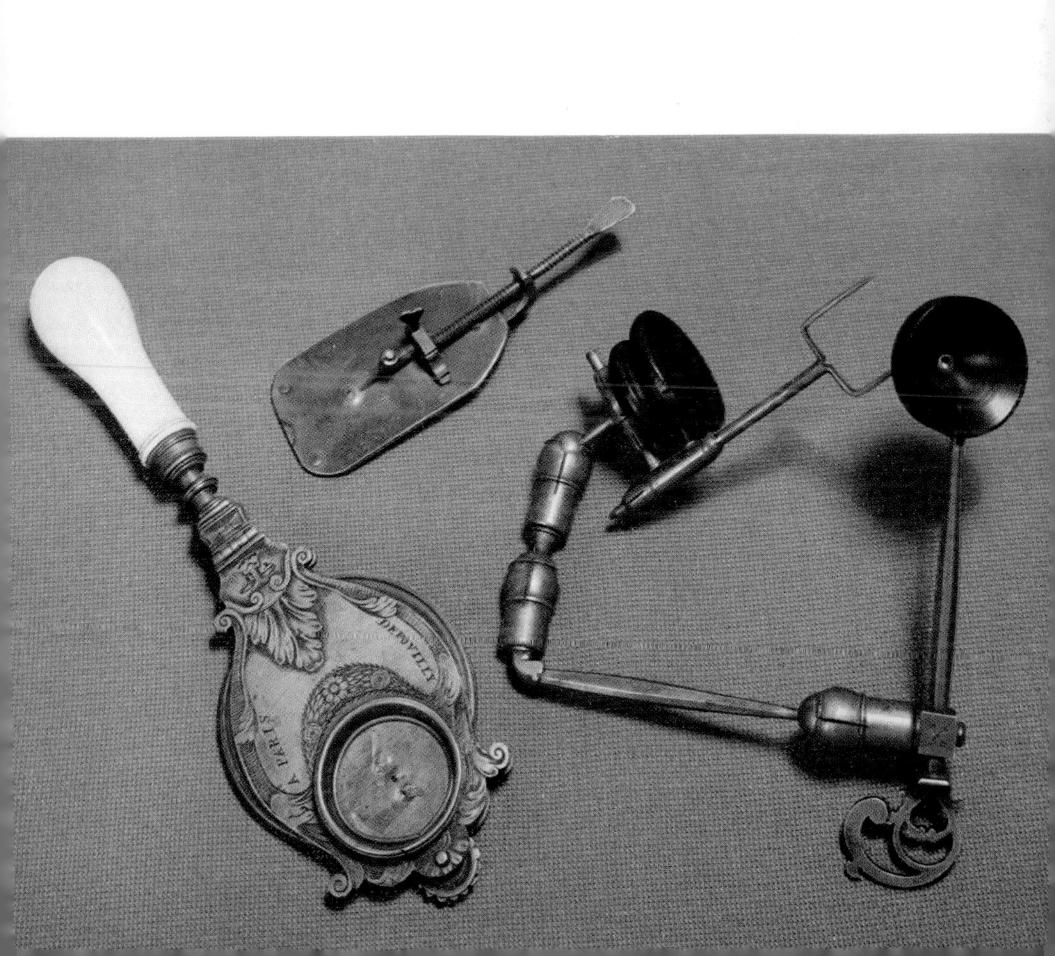
A PARIS

2 Compass and screw-barrel microscopes

The compass microscope originated at the end of the 17th century and became quite popular in the first half of the 18th century. It essentially consisted of a pair of compasses, one leg carrying the lens, and the other the object placed on the point. Focussing was performed by changing the angle between the legs.

The instrument in the photograph was probably made in the middle of the 18th century. The two arms of the compass are fixed to an ivory handle. The object can be either impaled by the point or held in the forceps as shown. The other arm carries the lens at the centre of a Lieberkühn. The distance between the two arms is adjusted by the link-screw.

James Wilson (c1665-1730) was a famous optical instrument maker of his period. In 1702 he described a type of microscope in an article in Philosophical Transactions of the Royal Society, a microscope which has since been commonly known as the Wilson Screw-barrel microscope.

The photograph shows one of his screw-barrel type microscopes, made in ivory, in the first quarter of the 18th century. The ivory barrel with the handle attached to it has at one end the viewing lens and at the other end a large ivory hollow screw which screws in and out of the other barrel. From one of the two open sides of the barrel one can see the two metal plates which carry the specimen slider, the metal plates being held against the ivory screw by the helical spring as seen in the photograph. Focussing is performed by this hollow screw which has a condensing lens at the far end.

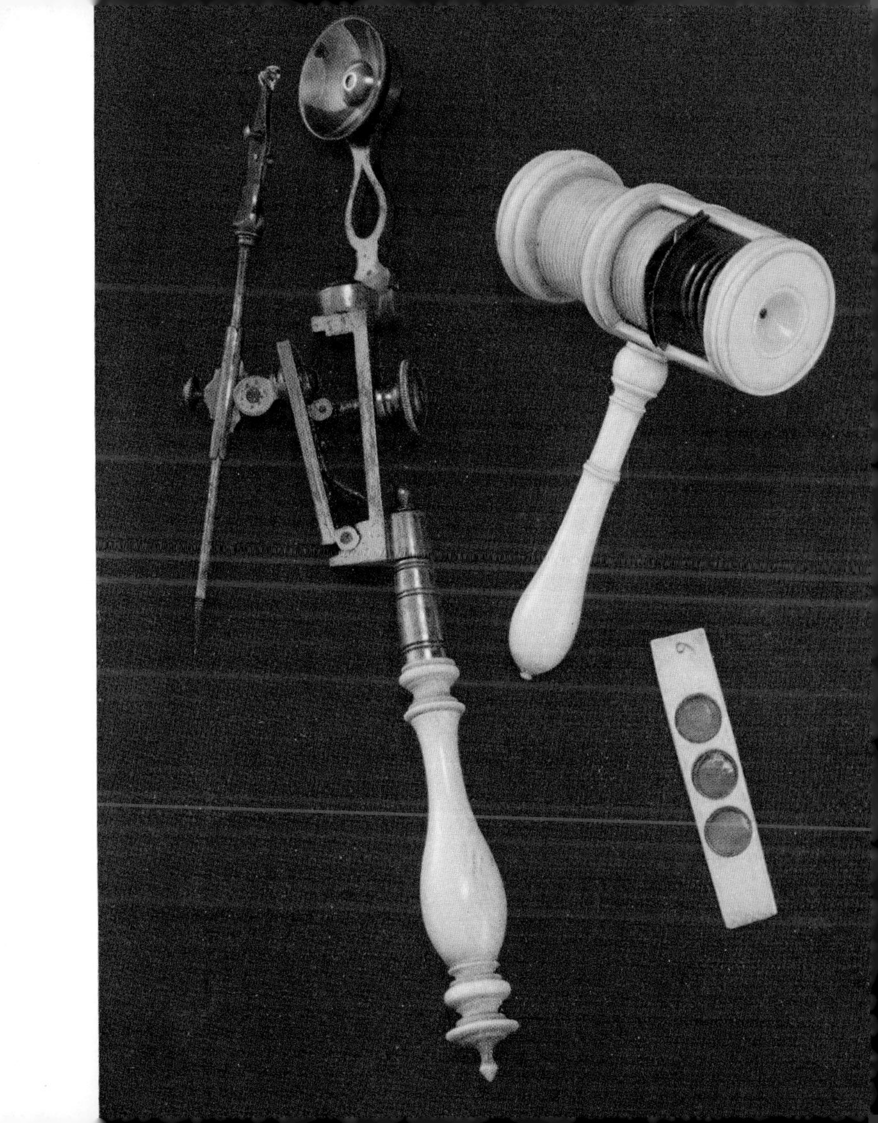

3 Hooke's microscope c1665

Robert Hooke was born in 1635 in the Isle of Wight. After a short apprenticeship to a portrait painter in London he attended the Westminster School and pursued his interest in Physics and Mathematics at Oxford. A versatile genius he has to his credit besides the well-known 'Hooke's Law', many other interesting devices such as a marine barometer, a spirit level, an air pump, an anchor escapement, spiral balance-spring, universal joint, etc.

In 1665 he published his monumental work *Micrographia* which established his name in the history of microscopy. This masterpiece on microscopy contains some of the most detailed illustrations of various objects he observed under the microscope.

The compound microscope in our photograph differs in many respects from the original described in his *Micrographia*. The main body of the microscope is about 300 mm long, made of cardboard, covered with dark red leather, ornamented with gold stampings. A bi-convex eye lens is enclosed in a wooden lens holder which has a dust-cap to house spare objectives. A plano-convex field lens is mounted inside the body of the tube. Objectives of various powers could be fitted on to the snout. Focussing is effected by rotating this threaded snout.

The whole microscope held by an arm, slides up and down a pillar mounted on a circular wooden (lignum vitae) base. The specimen can be mounted on a plate or a spike fixed to a rotating disc, below the objective.

The lamp and the condenser system shown in the photograph are a reconstruction from Hooke's drawings in his *Micrographia*. An original edition of his book is shown by the side of the microscope.

4 Early Italian compound microscopes

These three compound microscopes were made by Italian craftsmen probably in the latter half of the 17th century. The centre piece has engraved on it 'GIVSEPPE CAMPANI IN ROMA' the name of the famous Italian microscope maker (1635-95), a skilled optical craftsman of his time who worked in Bologna and Rome. The two other microscopes bear no inscription but are similar in design to the types of microscopes made by Campani and Divini (1620-1695).

Campani is considered to be the first to have developed the screw-barrel device as a means of focussing in compound microscopes. His microscopes were small in size and included a slide holder in the base of the instrument. This combination of small size and an attached slide holder enabled one to lift it to the sky to see transparent objects by transmitted light.

Divini appears to have been the first to introduce a doublet lens for the eye-piece.

The legs and collar of the microscope on the left are not original, being restorations performed at a later date. It has the usual bi-convex objective but has three lenses in the eye-piece tube a field lens and a doublet of bi-convex lenses.

The Campani instrument has two spirally threaded tubes, the top tube threading into the lower tube to vary the magnification, and the lower tube threading in the engraved collar to focus it on the object. The object slide is inserted between the base plate and the spring-loaded brass plate fixed above the base plate.

The instrument on the right is an earlier more crude form of a Campani type of microscope.

CAMPANI

5 An early compound microscope by Mann c1700

James Mann (c1660-1730) son of a tailor in Hertfordshire was apprenticed at an early age to Thomas King, a spectacle maker. He eventually became a master of the Spectacle Makers Company in 1716. It is interesting to note that he was fined by the Company in 1695 for making and selling 31 pairs of white spectacles ground only on one side and 5 pairs of black spectacles, 'being unlawfull and badd wares'. On his trade card he describes himself as an optician 'At the sign of Sir Isaac Newton, and two pairs of Golden Spectacles, near the West End of St Paul's, London, The Oldest Shoppe'.

This instrument is the earliest signed example of an English microscope in our collection. It is stamped in gold 'I. MANN. FECIT' on the vellum covering of the cardboard body and was made c1700. The basic form is a copy of the early continental style of Compound microscopes.

A tripod standing on a circular wooden base supports a threaded collar. The main body of the microscope screws in and out of this collar to focus on the specimen lying on the base. Optically this instrument is of interest since in this case the distance between the field-lens and the eye-lens can be varied by the draw tube, thus enabling a variation in the magnifying powers of the eye-piece system. The objective and the above lenses are bi-convex.

6 Marshall microscope 1715

John Marshall (1663-1725) was apprenticed to J. Dunning a turner and a telescope tube maker who was associated with Smethwick, a glass grinder making lenses for optical instruments. Marshall started work as a turner in Ivy Lane, London, where he had the opportunity of making his first microscopes for Sir Robert Boyle. In a trade card showing his place of business 'at the Sign of the Archimedes and two Golden Spectacles, in Ludgate Street, near St. Paul's Churchyard, London', he claims to be the first to have made it possible to observe 'circulation of Blood in Fishes'.

The instrument shown here bears the signature J. Marshall, London, and the date 1715 inside the drawer. The pillar about 300 mm long carrying the microscope is mounted on the large weighted wooden base with a ball and socket joint. This important joint not only enables the microscope to be tilted to a convenient angle for observations but also makes it possible for the main body to overhang the wooden base; thus facilitating the use of a candle underneath to examine specimens in transmitted light.

The stage holding the specimen slides is mounted on the same pillar thus keeping the object in focus when changing the inclination of the microscope. While all the previous compound microscopes had to be rotated bodily to bring the object in focus, Marshall was the first in this country to adopt the Helvelius' method, using two sleeves connected by a metal screw. One sleeve was fixed to the pillar while the other carried the microscope. Fine focussing was performed by turning an octagonal nut on this screw.

It has the conventional optical system of the period, namely an objective, a 'field-lens' and an eye-lens. Four of the original objectives still survive.

7 Culpeper microscope c1725

Edmund Culpeper (1660-1730) was apprenticed to Walter Hayes a maker of scientific instruments in Moorfields, London. He eventually took over this business and used the same sign of 'Cross Daggers'. Culpeper introduced (c1725) a simple design of the microscope stand which in various forms can be seen as the basis of microscopes manufactured in the next 100 years. It is essentially a tripod sitting on another tripod. His most important innovation was the introduction of a mirror (concave) mounted on gimbals beneath the stage. The mirror being concave not only helped to reflect the light into the microscope but also focussed it on the transparent specimen, thus doing away with the condenser of the Marshall type.

This early form of Culpeper microscope has a circular wooden base which carried a wooden stage on a tripod. Another smaller tripod on the stage supports a cylindrical cardboard tube covered in red vellum. The main body of the microscope slides into the red tube, the snout carrying the objective lens protruding below the tube. Focussing is effected by sliding the whole body up and down the red tube. The dust cap on the eye-lens has four compartments to house extra objectives of various powers ranging from x30 to x200. The optical system is the usual three lens system consisting of the objective, eye-lens and the field-lens. A condenser lens mounted on the stage helps to focus light on an opaque object.

8 Martin's drum microscope c1740

Benjamin Martin (1704-1782), an author of a large number of books on optics and mathematical instruments and a prolific maker of microscopes had some influence on the development of the microscope and other optical instruments of the period. He had the advantage over most of his contemporaries of having had a scientific training; for some time acting as an assistant to the famous scientific lecturer Desaguliers. He himself lectured on scientific topics and ran his own school at Chichester. An advertisement of the period stated 'Benjamin Martin, Chichester, Author and Teacher'. He started his career of an instrument maker at Chichester in 1738. Later he moved to Fleet Street in London. In 1780 he was joined by his son in partnership, and became 'B. Martin & Son'. A year later the firm was bankrupt.

The microscope proper consists of a green vellum covered cardboard tube with a large wooden mount for the eye-piece and a long wooden snout holding the objective. This complete compound microscope assembly slides into a cut away base tube which is covered in black fish skin on the outside. The lower part of this cut away outer tube is drum-shaped, one end of the drum being the wooden stage with the large hole in the centre. This drum is also cut out to let light fall on the plane mirror which is fixed at an angle of 45° to the wooden base. The top half of the base tube is cut away, obviously to throw light on opaque objects mounted on the stage. The microscope has the usual three lens optical system. A sliding brass dust-cap is fitted on top of the eye-piece.

9 The Cuff microscope c1744

John Cuff (1708-72) was apprenticed in 1722 to the well known microscope maker James Mann. He was elected a master of the Spectacle Makers Company in the years 1748-49. Having gone through the systematic training of the period for an optical instrument maker he opened his shop 'At the sign of the Reflecting Microscope and Spectacles, against Sergeant's Inn Gate in Fleet Street, London'. He had the patronage of the well known microscopist of the period Henry Baker (1698-1774) whose criticism and advice may have helped Cuff to modify and improve on the Marshall and Culpeper type microscopes. His outstanding contributions were the introduction of brass body and mounts, rack and pinion focussing and the design of a stage which enabled one to handle specimens freely without the Culpeper type legs getting in the way.

The advance on the previous instruments and the resemblance to the basic skeleton features of the microscopes of the succeeding centuries are obvious from the photograph. The firm rigid pillar stand and the large concave mirror are screwed into the wooden base which has a drawer full of accessories. The cross-shaped stage carries a condenser lens. The elegantly shaped brass body can be easily slid up and down the pillar. The objectives of various powers are numbered and the corresponding numbers are inscribed on the draw tube of the microscope and the pillar-stand. On screwing in a certain powered objective one needs only to set the draw tube at that number and slide and fix the microscope to the same number on the pillar. The fine focussing can now be made with the long screw behind the pillar.

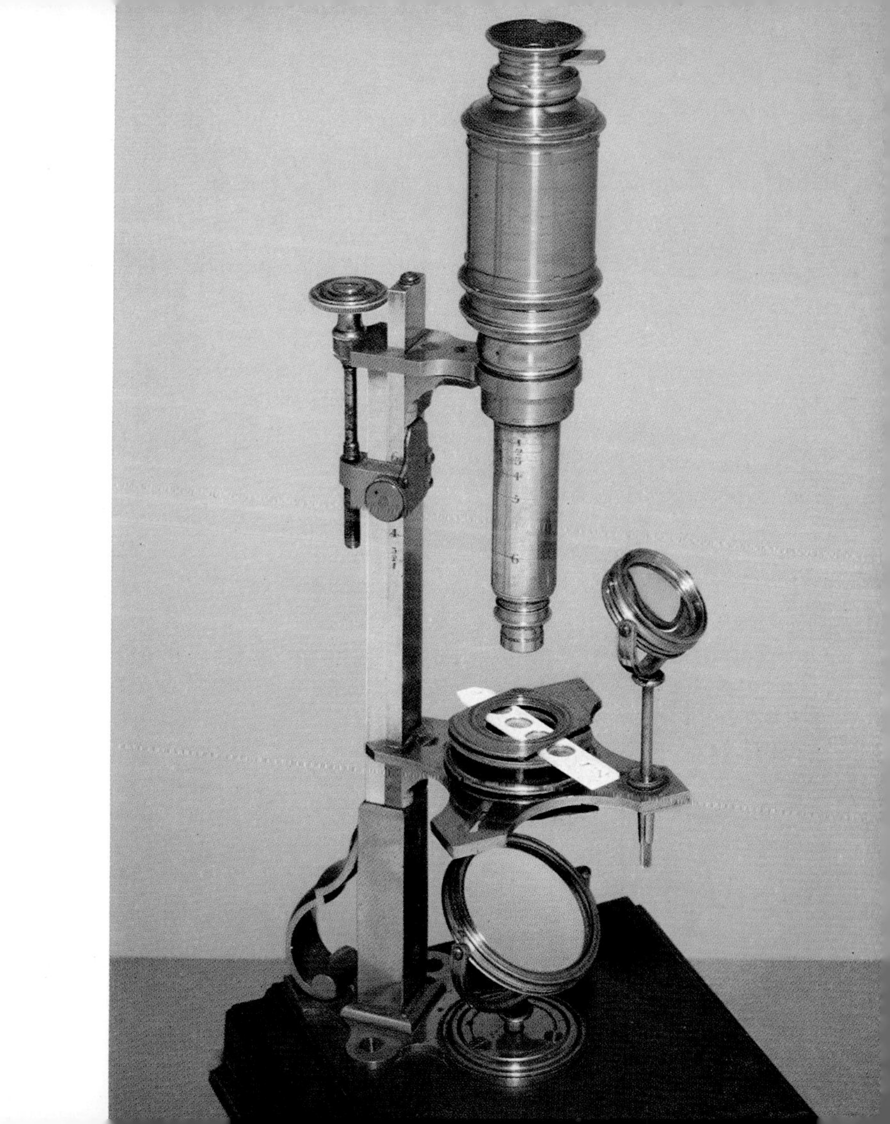

10 Adams' 'Prince of Wales' c1755

George Adams (1704-1773), an optician and instrument maker by trade was one of the most prolific master craftsman of his period ; he had a profound influence on the craftsmen of his time. One cannot help admiring his inventiveness and skill in the work he left behind. Very little is known of his early life. His trade cards show that he had established his shop at the sign of 'Tycho Brahe's Head, the corner of Racquet Court, in Fleet Street, London'. He was appointed an instrument maker to the East India Co. (1735-1736), to Royal Ordinance (1748-53) and to the Prince of Wales who was later to become his Royal patron as King George III. He had published many books, his *Micrographia Illustrata* (1747) having gone through several editions in his life-time.

This microscope was made for George III when he was still the Prince of Wales. No description can do justice to the detailed workmanship shown in this photograph. The whole microscope is hinged at its centre of gravity on trunions. This form of support gives it rigidity when tilted, a feature lacking in the previous instruments. The pillar carries two concentric wheels, one fixed and holding the microscope tube, the other rotating underneath the first wheel and carrying eight holders for the different objectives. Focussing is performed by sliding two rectangular bars on each other, one carrying the microscope and the other the mirror and the stage. The stage is also interesting since it has micrometer screws registering in two directions at right angles, capable of reading to ten thousandth of an inch. It has the usual three lens optical system.

11 Solar microscope by Martin c1760

Benjamin Martin made several types of microscopes in his time. In about the middle of the 18th century he sold 'cabinets of optical instruments' which contained various instruments packed together in a case very neatly arranged each in its own compartment; for instance one contained a terrestrial telescope, a drum microscope, a scioptric ball and socket with suitable lenses and a solar microscope.

The solar microscope shown here is an item from one of such cabinets and is signed 'B. Martin, London.' With this instrument fitted into a hole in the window-shutter, it is possible to project a large image on a screen thus enabling several people to look at a specimen in the microscope. Sunlight was used as the powerful source of light, hence the name 'Solar' microscope.

A large mirror which reflects sunlight into the microscope and is capable of adjustment for tilt and rotation is fixed on one side of a 136 mm square wooden frame. On the other side of this frame is a cardboard tube covered with green shark-skin and carrying a condenser lens. A draw-tube carries the Wilson-type microscope, shown here with a specimen slide fitted into the spring-stage. A metal slider, fitted with a choice of three lenses of different powers, is inserted at the eye-piece end of this microscope. The brass tube holding this lens slider is moved to and fro by the rack and pinion to focus the object on the screen. Sunlight is focussed on the specimen by sliding the cardboard draw-tube.

N.°5

12 Adams' silver microscope c1770

This is one of the masterpieces of craftmanship by the well known instrument maker George Adams. It was probably made for George III. The microscope had remained in the possession of the Royal Family until it was presented to the Science Museum by George VI. The photograph does full justice to the central Corinthian pillar on a pedestal, the short pedestals on either side bearing ornamental urns connected together by decorated silver tubing, the foliage, the cherubim figures, the two partially draped female figures holding the microscope tube, all fine specimens of the silversmith's art; the most artistically and elaborately decorated instrument ever constructed.

However it is not one of Adams's best instruments for serious scientific work. It stands 736 mm high and is cumbersome to use.

13 A 'museum microscope' c1810

Thomas Winter an optician and dealer in all kinds of instruments was in business in 1800 at 37 Brewer Street, Golden Square, London. He is credited to have been the first to have put this type of microscope on the market.

The salient feature of this instrument is the large ivory stage, a revolving cylinder carrying three rows of apertures sixty-five in all. The microscope swivels on the stand to enable one to focus it on various specimens over the width of the cylinder. An extremely useful microscope used for teaching purposes and for exhibiting to a large group of people, a set of pre-selected specimens. The circular brass base is inscribed T. Winter, 9, New Bond Street, London.

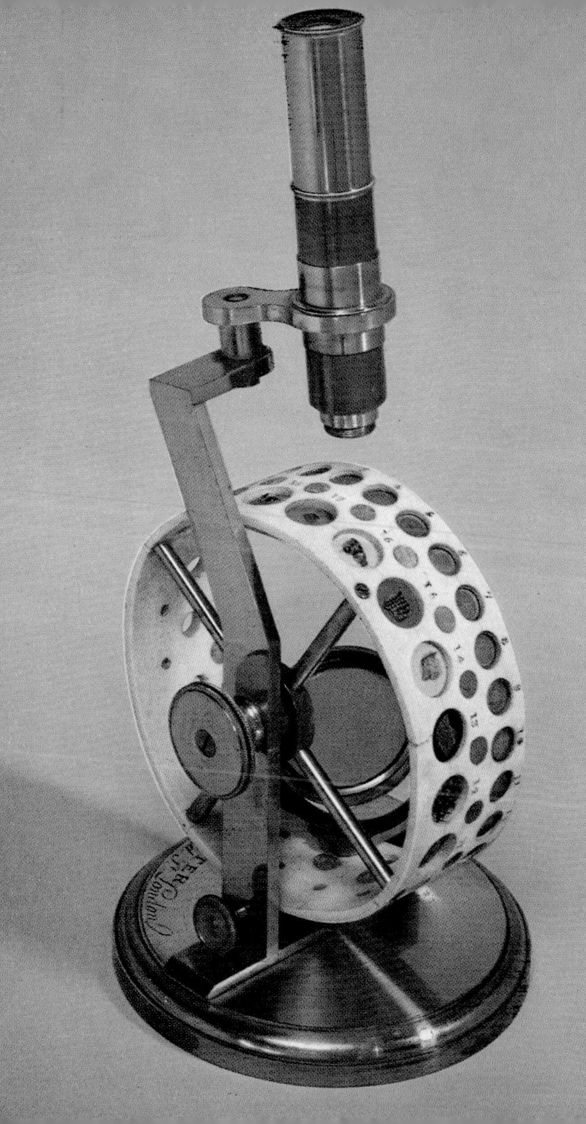

14 A Dollond microscope

John Dollond was born of humble parents who had come to England in 1685 from Normandy and had settled down as silk-weavers in Spitalfields, London. Young John (1706-1761) spent most of his time when away from the loom, studying astronomy, mathematics and optics. He carried on his business as a silk-weaver while taking intense interest in the study of optics. By the middle of the 18th century he was known to the scientific circles of London for his wide knowledge of theoretical optics. In 1752 he joined his son Peter (1730-1820) in business at 'the Sign of the Golden Spectacles and Sea Quadrant, near Exeter Change in the Strand'. In 1759 Dollond's were the first to introduce on the market an achromatic combination of lenses for telescopes and took out a patent for the method of making this combination.

It took the optical industry another quarter of a century to acquire sufficient technical skill in grinding and polishing very small lenses to make an achromatic combination for the tiny objective of a microscope. The first achromatic microscope objective has been attributed to an amateur, a colonel in the Amsterdam cavalry, Francois Beeldsnyder (1755-1808). Most of the well-known opticians in Europe and England were soon made to realise its importance for the experimental microscopist and by the middle of the 19th century all the better type of microscopes had an achromatic objective.

The 600 mm high microscope in the photograph has inscribed on one of the legs, 'Dollond, London.' It has two noteworthy features, the oil lamp mounted on gimbals remaining vertical for all inclinations of the microscope and the completely enclosed, horizontal stage movements. The objective is an achromatic combination. This microscope was probably made in the early 19th century.

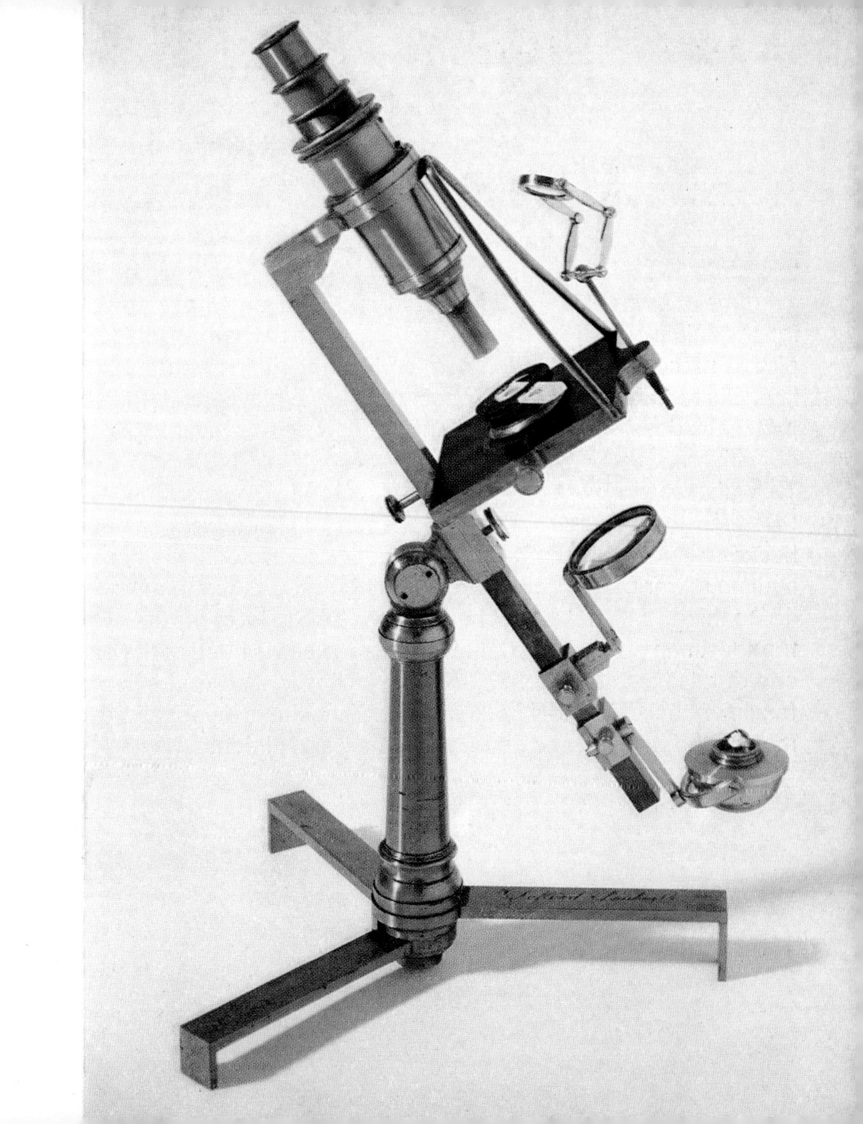

15 Amici's reflecting microscope c1830

Professor G. B. Amici (1785-1863) an Italian experimental physicist was keenly interested in the design of optical instruments. He was worried by the inherent spherical aberrations introduced by the achromatic combinations of the microscope objectives of the first quarter of the 19th century. After many initial unsuccessful attempts to overcome this problem he gave up the idea of making an ideal achromatic combination with lenses, and used an elliptical mirror for the initial magnification in a microscope: when an object is placed near one of the focii of the ellipsoidal mirror an enlarged image is formed at the other focus. Amici also played an important part in the development of the immersion lenses for the microscope objectives.

This microscope was made for Dr Wollaston by Amici who was at that time the Director of the Observatory at Florence. The long horizontal brass tube contains the concave ellipsoidal reflector and a plane mirror on the axis of this reflector, mounted at an angle of 45° to this axis. A hole is cut in this brass tube to let in the light from the specimen on the stage below. An eye-piece is fitted at the other end of this brass tube to magnify the image formed at the second focus of the ellipsoidal reflector. The specimen is focussed by moving the stage up and down with a rack and pinion.

16 A Cary microscope

William Cary (1759-1825) served his apprenticeship under Ramsden the famous optician and instrument maker of this period. He set up his first business at 272, Strand, London in 1790. Later he moved to 182, Strand, next door to his brother John who had his own business as a globe-maker and map-engraver. Both shops were unfortunately completely destroyed by fire in 1820.

This all-brass microscope 500 mm high has 'Cary, London' inscribed on one foot. A feature worth noting is that the whole instrument can be folded into one plane and packed flat into a specially constructed case: the feet pivoted on the central pillar fold into each other; the mirror, the condenser, the stage and the arm carrying the microscope swivel into position. The microscope body unscrews at the narrow end of the snout.

The stage has a rack and pinion for focussing and the microscope is positioned on the specimen by the swivel and the rack and pinion on the arm. The optical system has an objective, a field lens and a triple system of lenses for the eye-piece. This triple system consists of a bi-convex doublet and a plano-convex single lens near the eye.

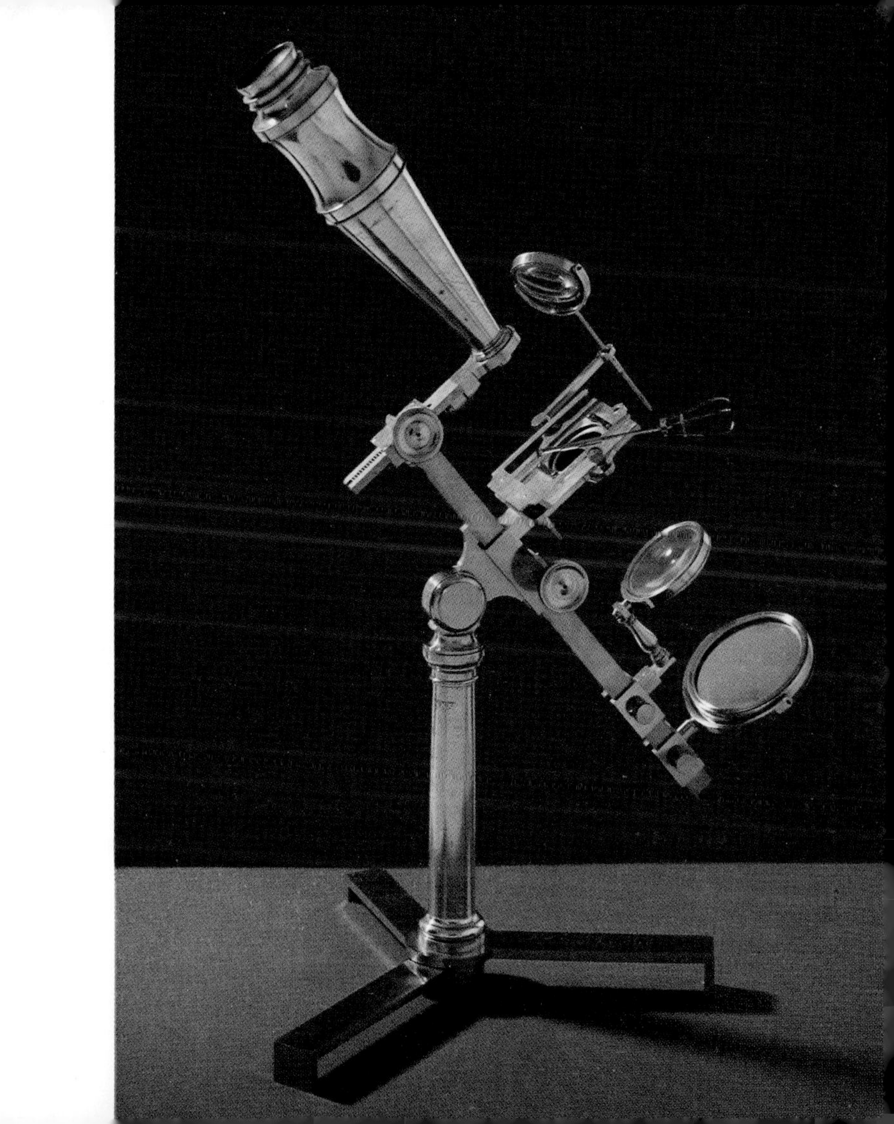

17 'Varley stirrup-lever' microscope

Cornelius Varley (1781-1873) a painter and inventor who lived at Clarendon Square, Somers Town, was one of the founder members of the Microscopical Society of London in 1839, now known as the Royal Microscopical Society. Varley demonstrated to the Microscopical Society in 1841 his invention of the 'lever stage movement' which enabled the stage to be moved laterally in any direction, thus keeping living aquatic objects under continuous observation. Hugh Powell made the microscopes for Varley.

In the 1830's Hugh Powell (1799-1883) was making microscopes for the trade at 24 Clarendon Street, Somers Town, London. His brother-in-law Lealand joined in partnership in 1841 and since that date the firm's signature 'Powell and Lealand' has appeared on some of the most beautiful refined and precise instruments for high resolution microscopy. The business was carried on up to 1914 by Hugh's son Thomas who had taken over after the death of his father.

The photograph shows a Powell and Lealand's improved form of the microscope devised by Varley, inscribed on the base 'Powell and Lealand, London'. The stirrup-lever for easy manoeuvring of the top part of the stage and the swivelling forceps for holding objects under the microscope are clearly seen in the photograph. The microscope has the usual three lens optical system, the objective being achromatic. It stands only 330 mm high. The stage is mounted on the vertical hollow stand and is moved for coarse focussing by rack and pinion. The large milled-head screw at the lower end of the hollow stand pushes the body-supporting arm up and down for fine focussing.

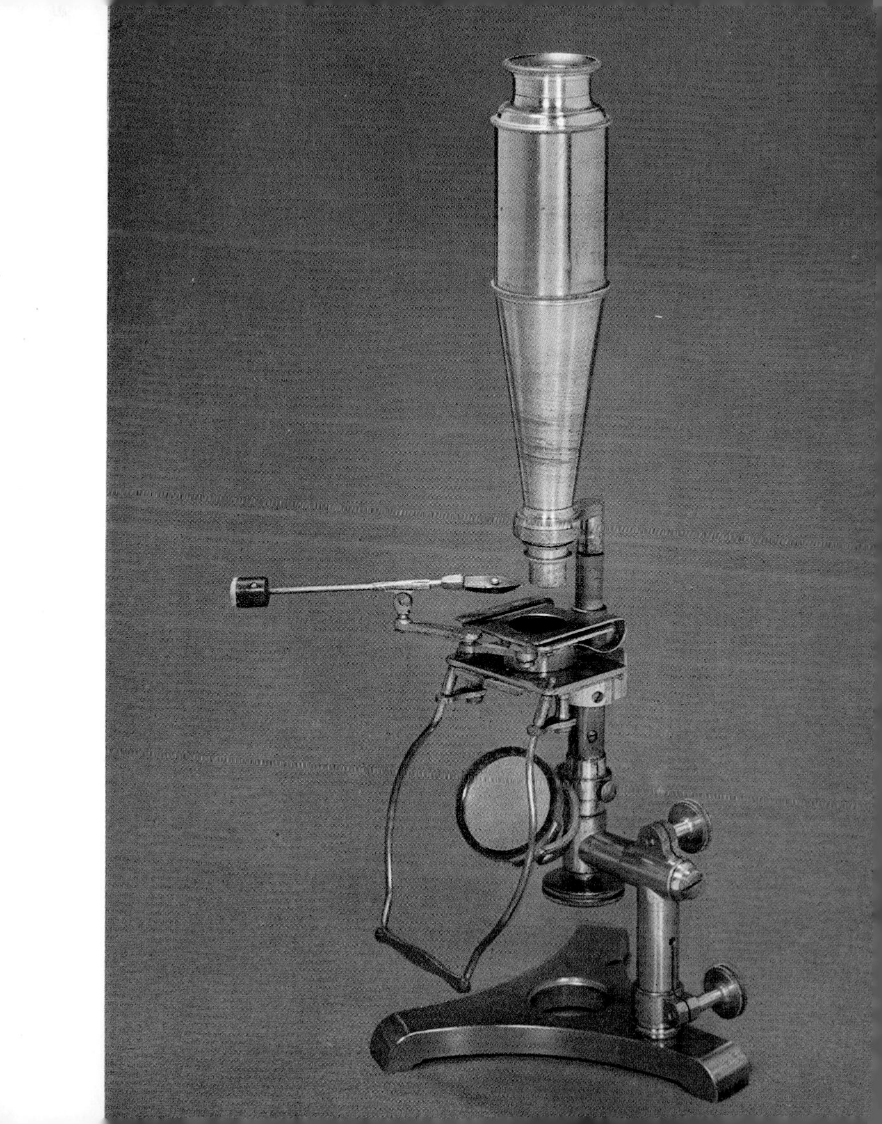

18 Wenham-Ross Binocular microscope c1860

The Capuchin friar, le Père Cherubin, of Orleans wrote in 1677, 'Some years ago I resolved to effect what I had long before premeditated, to make a microscope to see the smallest objects with the two eyes conjointly . . .' No interest was shown in binocular microscopes for the next hundred and fifty years till in 1838 Wheatstone published his paper on Stereoscopes and binocular vision. An American Professor Riddel of New Orleans is credited to have been the first to have made a useful binocular instrument in 1851.

Wenham, a man with great practical knowledge of the design and construction of microscopes and who was working for the firm of Ross, designed a prism which formed the basis of this binocular instrument. The prism, mounted to cover one half of the objective lens, deflected the rays into the second ocular tube fixed at an angle to the usual vertical tube. The uncovered half of the objective formed the normal image in this vertical tube. These two images gave the binocular vision when viewed through the two eyepieces.

This 570 mm high microscope bears the inscription 'Ross, London', and their stamp of precision construction and attention to detail. The stage has provision for rotational movement besides the usual two-directional traverse. The substage carrying the condenser system also has a rotational movement besides the up and down traverse along the main limb. There is provision to mount two achromatic objectives, either of which can be swivelled into position. At the eye-piece end the binocular bodies are linked by a rod which racks the two eye-pieces simultaneously in and out, thus separating the distance between the eye-pieces to help adjust to an observer's binocular vision.

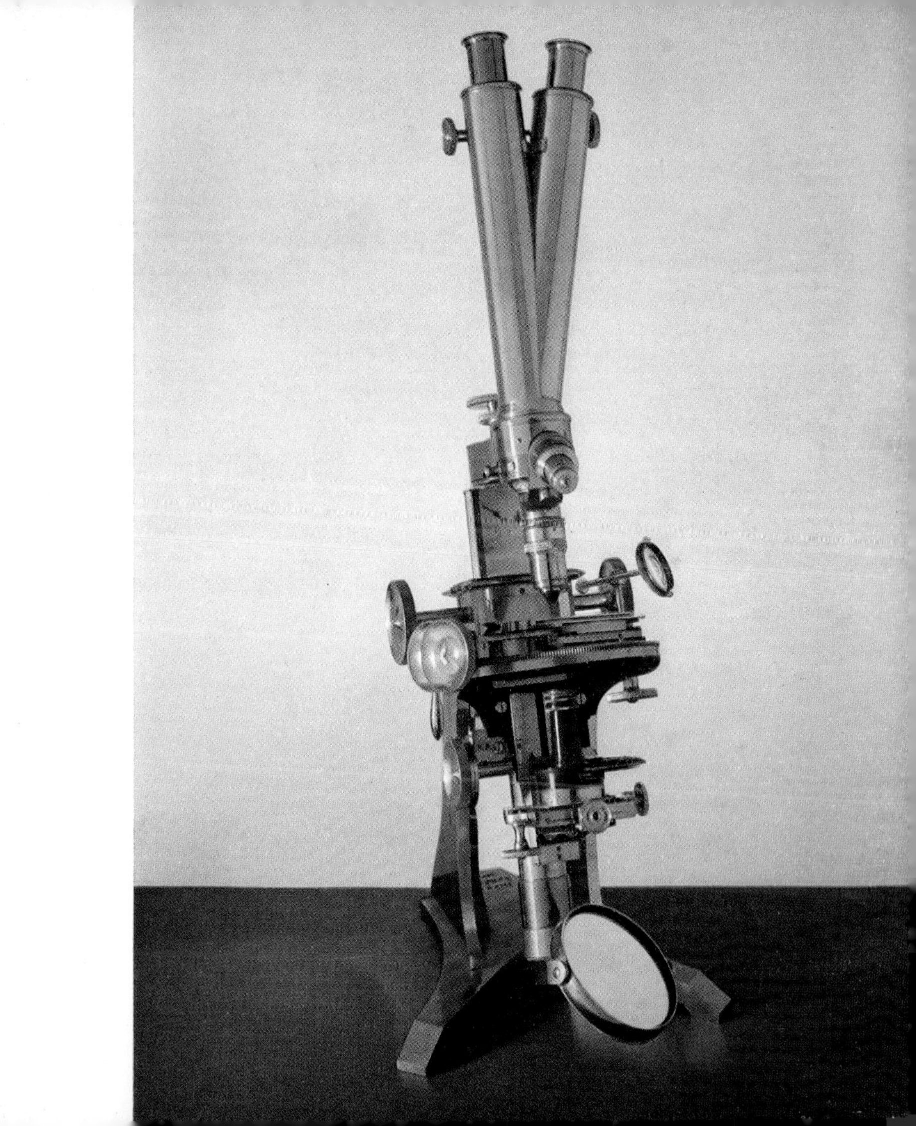

19 Thury's multi-ocular microscope

This unusual instrument was made by the Geneva Optical Co., probably in the last quarter of the 19th century for Professor M. Thury a Swiss physicist. It was presumably designed to save time when demonstrating to a group of students. In 1861 Professor Thury helped Professor Aug. de la Rive of Geneva to start the first factory in Switzerland which concentrated exclusively on the construction of 'physical and mechanical instruments'.

This microscope is provided with five eye-piece tubes mounted on a central box which contains a 90° prism. The prism can be readily rotated to direct light from the single objective into any one of the eye-pieces, thus enabling five different observers to view in quick succession, but not simultaneously, any specimen under observation. Four of the tubes are provided with rack adjustment for the eye-piece. The demonstrator using the central eye-piece focusses the microscope and the students adjust their eye-pieces to suit individual vision. It is provided with a substage condenser system on rack and pinion.

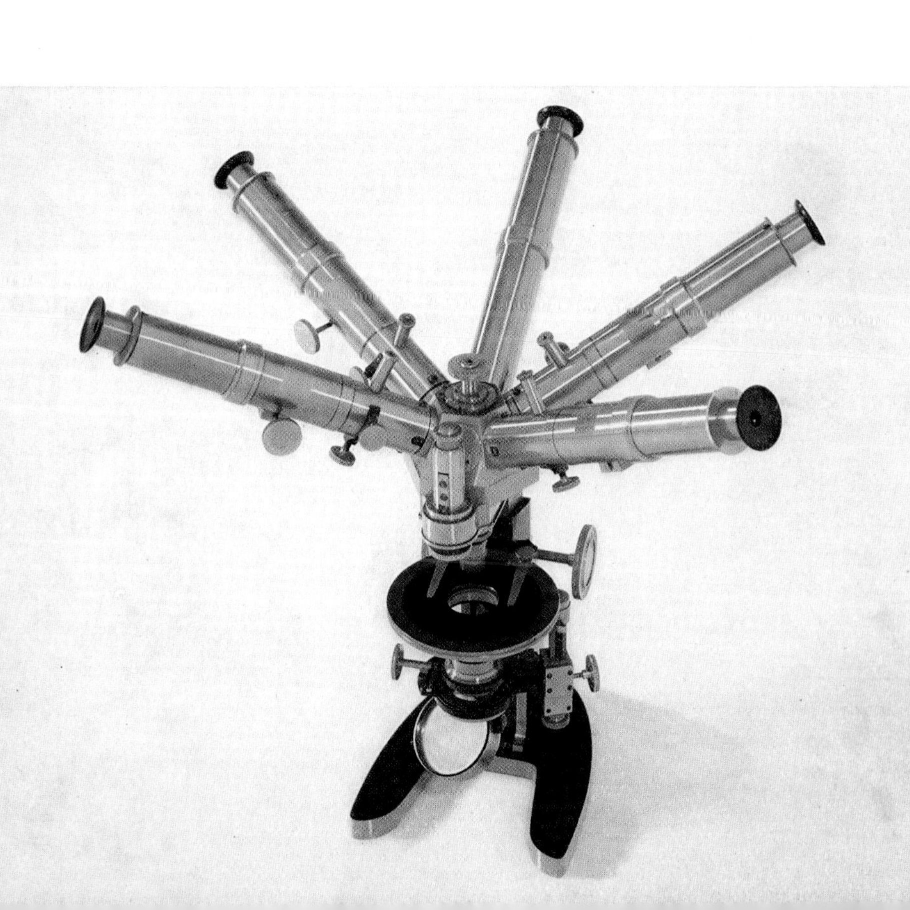

20 Nachet's inverted microscope c1885

Camille Sebastien Nachet (1799-1881) after a period of military service and six years experience of lens manufacture with the famous French instrument maker Vincent Chevalier, opened his own business in 1840. He became well known in the French scientific circles for the various types of microscopes and the powerful achromatic objectives he made. His son Jean-Alfred Nachet (1831-1908) succeeded his father and kept up the reputation of the firm till he retired in 1899.

The instrument in the photograph is 300 mm high and has inscribed on it 'Nachet et Fils, 17, Rue St. Séverin, Paris'. The circular base has four pillars supporting a platform which carries a box containing a front silvered mirror. Light from the stage above, passes through the objective and is reflected into the eye-piece tube by the inclined mirror. The initial focussing is performed by sliding the objective tube while the fine focussing is made by the screw between the pillars moving the stage up and down. On the top of the stage one can see the housing for a plane mirror and a condenser lens.

This inverted chemical microscope was made 'for the purpose of viewing objects from their underside when heat or reagents are being applied to them. It meets the requirements of observers engaged in the 'cultivation' of the minute organisms which act as ferments'.

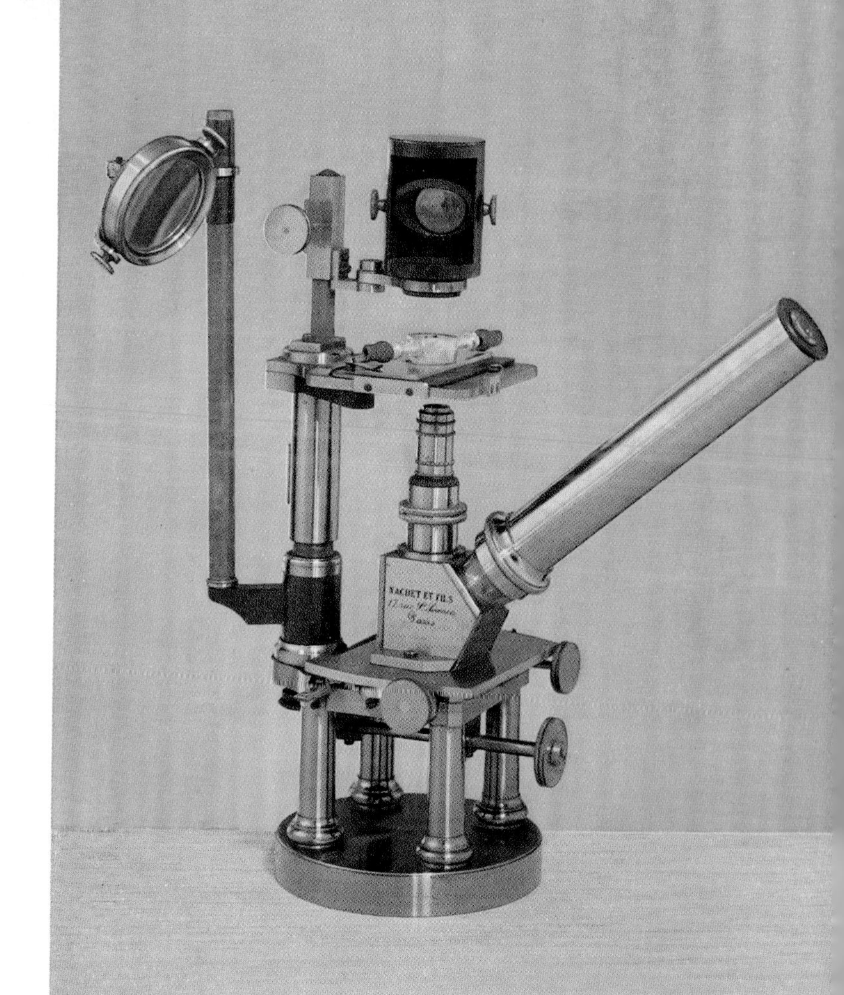
NACHET ET FILS

Science Museum illustrated booklets

Other titles in the series cover such subjects as Aeronautics, Road and Rail Transport, Chemistry, Timekeepers, Ship Models, etc., and can be obtained from the Government Bookshops listed on cover page iv (post orders to PO Box 569 London SE1)

35p each (by post 37½p)

Printed in England for Her Majesty's Stationery Office by W. Heffer & Sons Ltd
Cambridge Dd. 502170 K80